DANGEROUS JOBS

SMOKE JUMPERS

CLARA CELLA

Lerner Publications ◆ Minneapolis

Lerner Publications Company
An imprint of Lerner Publishing Group, Inc.
241 First Avenue North
Minneapolis, MN 55401 USA

For reading levels and more information, look up this title at www.lernerbooks.com.

Main body text set in ITC Franklin Gothic Std.
Typeface provided by Adobe Systems.

Library of Congress Cataloging-in-Publication Data

Names: Cella, Clara, author.
Title: Smoke jumpers / Clara Cella.
Description: Minneapolis : Lerner Publications, [2023] | Series: Updog books. Dangerous jobs | Includes bibliographical references and index. | Audience: Ages 8–11 | Audience: Grades 4–6 | Summary: "Smoke jumpers jump out of aircraft to fight wildfires in remote locations. Learn about their gear, their techniques, and the dangers these firefighting heroes face"— Provided by publisher.
Identifiers: LCCN 2022015375 (print) | LCCN 2022015376 (ebook) | ISBN 9781728475561 (lib. bdg.) | ISBN 9781728486239 (pbk.) | ISBN 9781728481944 (eb pdf)
Subjects: LCSH: Smokejumpers—Juvenile literature.
Classification: LCC SD421.23 .C45 2023 (print) | LCC SD421.23 (ebook) | DDC 634.9/618—dc23/eng/20220624

LC record available at https://lccn.loc.gov/2022015375
LC ebook record available at https://lccn.loc.gov/2022015376

Manufactured in the United States of America
1 – CG – 12/15/22

Table of Contents

FACE THE FIRE

The smoke jumper hangs in the air. They see the wildfire below. That's where they need to be.

wildfire: a fire that moves quickly through forests and grasslands

Smoke jumpers fight wildfires. They parachute to the ground from airplanes.

parachute: to jump from a great height strapped to a large cloth that slows one's fall through the air

They use saws and other tools
to clear away trees and grasses.
Wildfires can't spread without this
plant material.

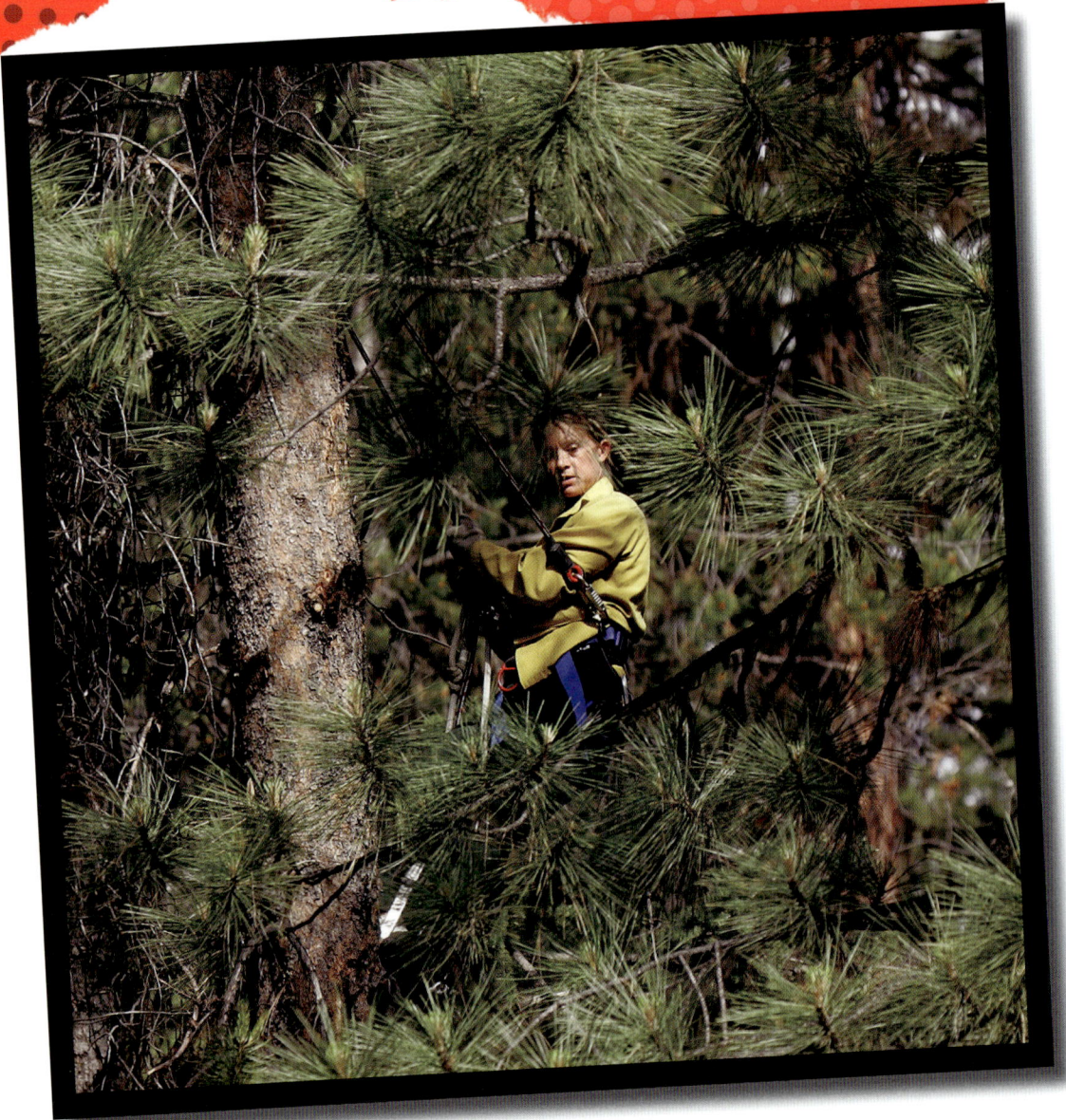

Smoke jumpers face many dangers. Their parachute might get caught in a tree.

They may get burned or break bones. Breathing in smoke can make them sick.

UP NEXT! FIRST JUMPS.

DROP IN, FIGHT HARD

The first US smoke jumpers started working in 1940. Before then, firefighters on the ground struggled to reach remote areas. Smoke jumpers could get there faster.

remote: faraway and hard to reach

GEAR CLOSE-UP

Smoke jumpers wear special gear to keep them safe.

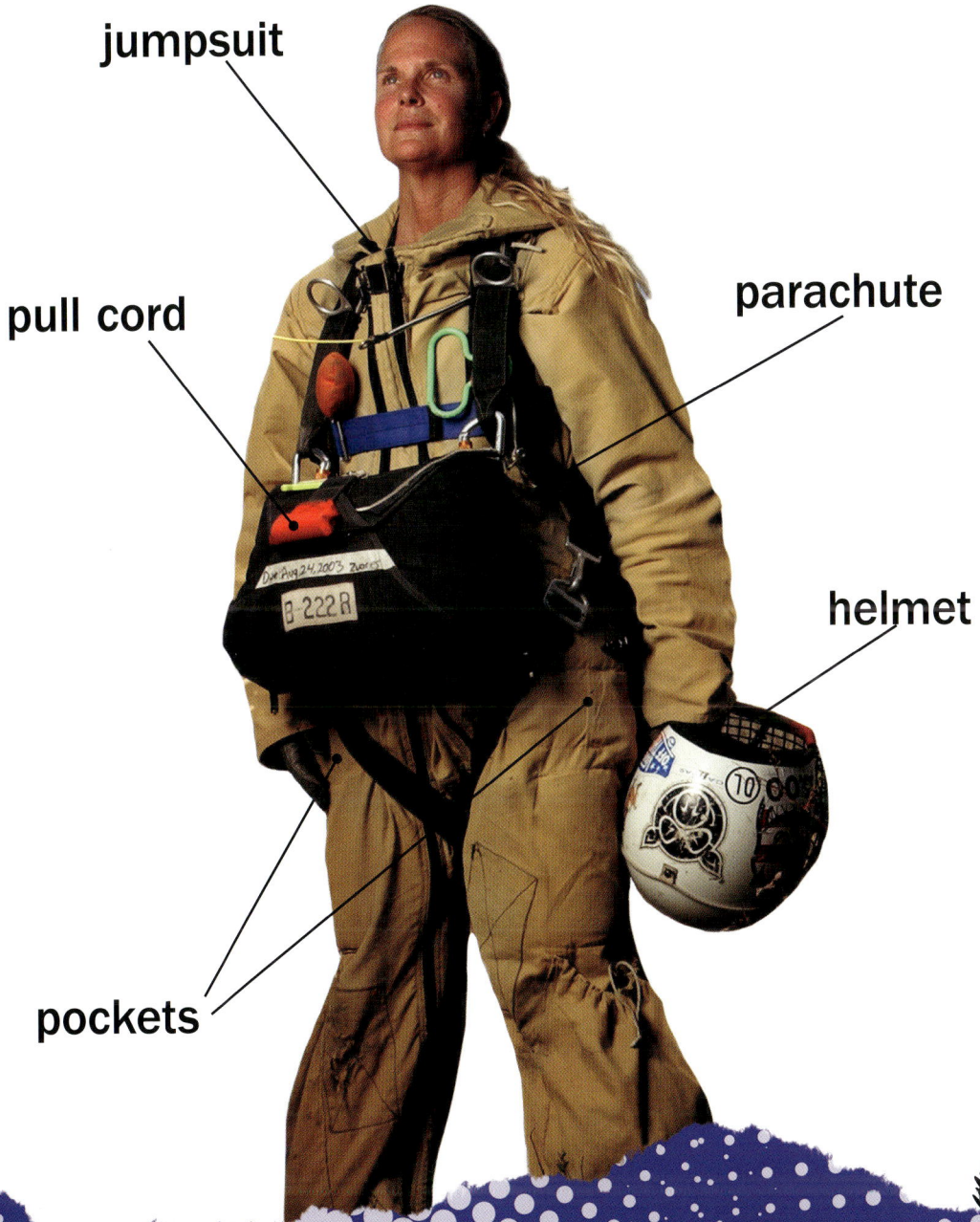

SMOKE JUMPER

jumpsuit

pull cord

parachute

helmet

pockets

Smoke jumpers are top firefighters.
They train to hike over rough
ground.

Each smoke jumper carries a gear pack weighing up to 115 pounds (52 kg).

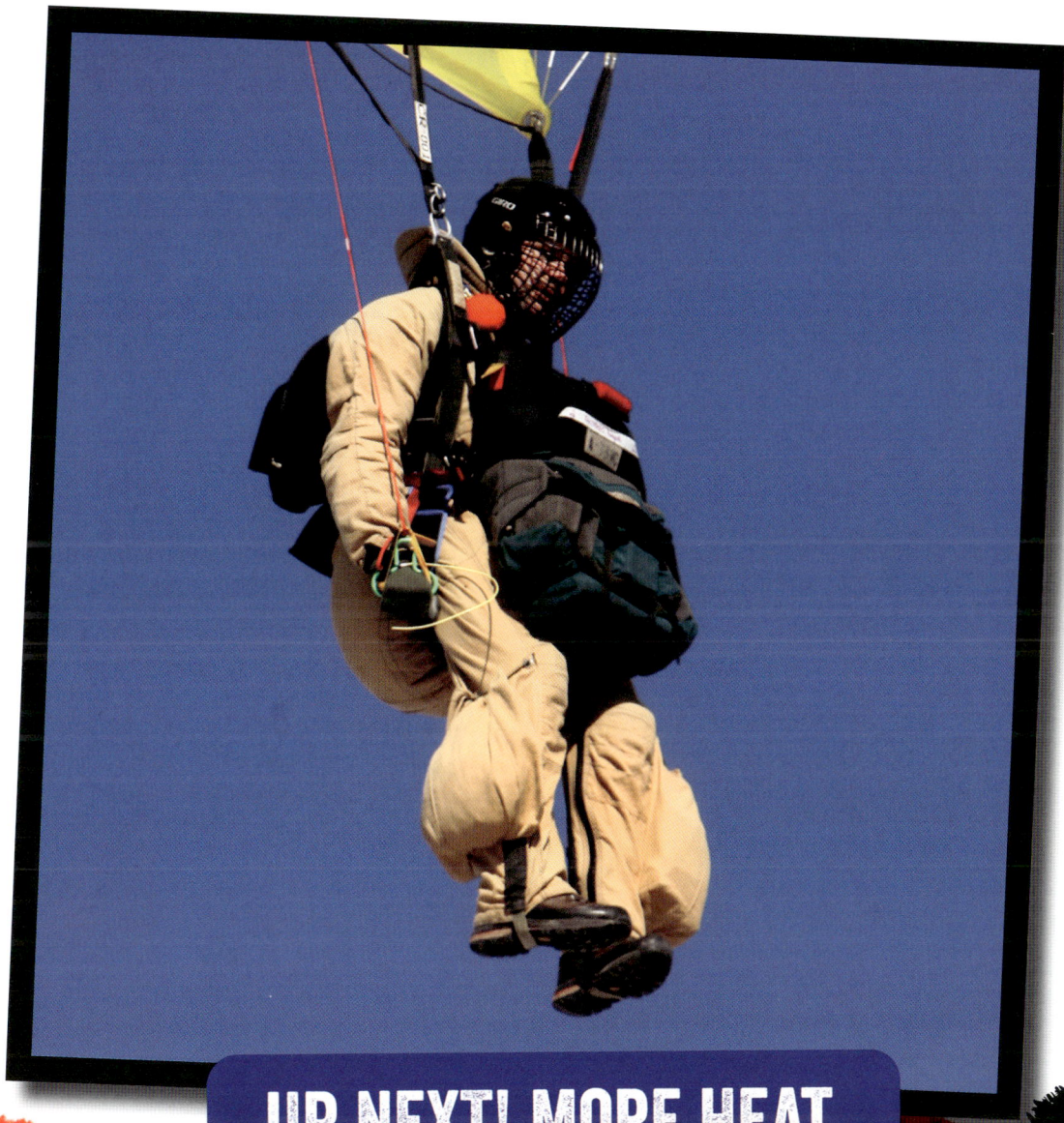

UP NEXT! MORE HEAT.

15

SUIT UP, STAY SAFE

Smoke jumpers wear special jumpsuits and helmets.

The padded suits do not easily rip or burn. Smoke jumpers also carry extra parachutes.

Climate change is causing more wildfires than ever.

climate change: a long-lasting shift in Earth's weather patterns

Smoke jumpers are working longer hours. They work as a team to get the job done.

Smoke jumpers have dangerous jobs. But they help keep people and property safe.

property: land or objects owned by someone

MEET A SMOKE JUMPER

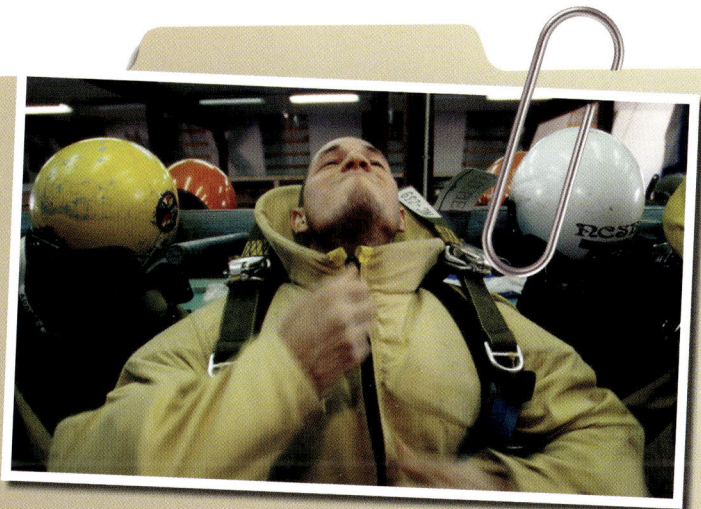

NAME: Jason Ramos

BACKGROUND: Ramos grew up in southern California. He became a firefighter at the age of 17.

CLAIM TO FAME: Ramos started smoke jumping in Washington in 1999. In 2015, he wrote about his life in *Smokejumper: A Memoir by One of America's Most Select Airborne Firefighters*. He is one of fewer than 6,000 smoke jumpers in US history.

Glossary

climate change: a long-lasting shift in Earth's weather patterns

parachute: to jump from a great height strapped to a large cloth that slows one's fall through the air

property: land or objects owned by someone

remote: faraway and hard to reach

wildfire: a fire that moves quickly through forests and grasslands

Check It Out!

Ducksters: Earth Science for Kids—Forest Fires
https://www.ducksters.com/science/earth_science/forest
_fires.php

Kiddle: Parachuting Facts for Kids
https://kids.kiddle.co/Parachuting

Murray, Julie. *Smokejumpers*. Minneapolis: Abdo Zoom, 2021.

Potenza, Alessandra. *All About Wildfires*. New York: Children's Press, 2021.

Smokey for Kids
https://smokeybear.com/en/smokey-for-kids

Westmark, Jon. *Smoke Jumpers in Action*. Mankato, MN: Child's World, 2017.

Index

Photo Acknowledgements

Image credits: Justin Sullivan/Staff/Getty Images, p.4; Straight 8 Photography/Shutterstock, p.5; Chris Lombardi/Getty Images, p.6; Justin Sullivan/Staff/Getty Images, p.7; Justin Sullivan/Staff/Getty Images, p.8; Jorge Enguidanos/EyeEm/Getty Images, p.9; Tillman/Wikimedia, p.10; David Ryder/Polaris/Newscom, p.11; David Ryder/Polaris/Newscom, p.12; Chris Butler/ZUMAPRESS/Newscom, p.13; David Ryder/Polaris/Newscom, p.14; Eric Engman/News-Miner/ZUMA Press/Newscom, p.15; Lance Cheung/ZUMA Press/Newscom, p.16; Ed Lallo/ZUMA Press/Newscom, p.17; Gaylon Wampler/Getty Images, p.18; Justin Sullivan/Staff/Getty Images, p.19; Tim Matsui/Stringer/Getty Images, p.20; Tim Matsui/Stringer/Getty Images, p.21; Mega Pixel/Shutterstock, p.22;

Design element: Infostocker/Getty Images

Cover: Patrick Orton/Getty Images; Gorgev/Shutterstock